SMALL BABIES

J. McQUADE

For my dear sister Yvonne and brother-in-law Duncan,
with love,
Jackie.

10 9 8 7 6 5 4 3 2 1
First published in the United Kingdom in 2000
by David Bennett Books Limited, United Kingdom.

Distributed by Sterling Publishing Company, Inc. 387, Park Avenue South, New York, N.Y. 10016.
Distributed in Canada by Sterling Publishing, c/o Canadian Manda Group,
One Atlantic Avenue, Suite 105, Toronto, Ontario, Canada M6K 3E7.

ISBN 0-8069-7541-5
Printed in China.

Small Babies

Jacqueline
McQuade

Baby
Mice

A baby mouse does not have a special name.
These tiny baby mice are sharing a juicy piece of apple.

Otter
Cub

A baby otter is called a cub.
This little otter cub stays close to the riverbank
when she goes swimming on her own.

Baby
Bunnies

A baby rabbit should be called a kitten
but the word 'bunny' is more popular.
*These baby bunnies are brother and sister
and they love to cuddle up to each other.*

Raccoon
Cub

A baby raccoon is called a cub.
This raccoon cub loves to spend his time climbing trees.

Baby Koala

A baby koala does not have a special name.
*This cute little koala clings onto his mommy's back
and just enjoys the ride!*

Owlets

A baby owl is called an owlet.
*These fluffy owlets are waiting for their mommy
to bring them something to eat.*

Baby
Guinea Pigs

A baby guinea pig does not have a special name.
These baby guinea pigs can be quite mischievous –
just look what they have done to these balls of wool!

Baby Squirrels

A baby squirrel does not have a special name.
These little squirrels are already old enough to climb trees
and they love to play hide-and-seek.